U0392465

Education
248

软体机器人

Soft Robotics

Gunter Pauli

[比] 冈特·鲍利 著

[哥伦] 凯瑟琳娜·巴赫 绘

何家振 译

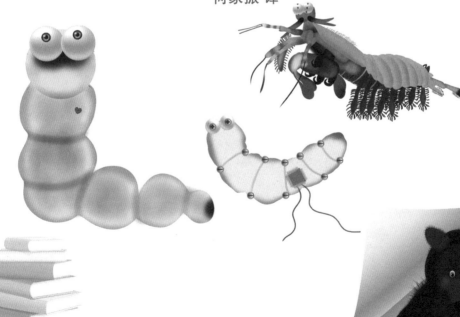

上海远东出版社

丛书编委会

主　任：贾　峰

副主任：何家振　闫世东　郑立明

委　员：李原原　祝真旭　牛玲娟　梁雅丽　任泽林

　　　　王　峃　陈　卫　郑循如　吴建民　彭　勇

　　　　王梦雨　戴　虹　靳增江　孟　蝶　崔晓晓

特别感谢以下热心人士对童书工作的支持：

匡志强　方　芳　宋小华　解　东　厉　云　李　婧

刘　丹　熊彩虹　罗淑怡　旷　婉　杨　荣　刘学振

何圣霖　王必斗　潘林平　熊志强　廖清州　谭燕宁

王　征　白　纯　张林霞　寿颖慧　罗　佳　傅　俊

胡海朋　白永喆　韦小宏　李　杰　欧　亮

目录

Contents

一只蛆注视着螳螂虾表演她的经典绝技：重拳击倒她的猎物。

　　"上勾拳，漂亮！"

　　"这可不是上勾拳，这只是普通一击。我通常会把我的食物击碎，即使我失手，猎物也会被打晕，从而更容易被抓到。"

A maggot is watching a mantis shrimp doing what she is famous for: knocking out her prey with a powerful punch.

"Good uppercut!"

"That wasn't an uppercut, it was just a blow. I usually smash my food to pieces, but even when I miss, my prey still gets stunned and is easy to catch."

... watching a mantis shrimp ...

产生热气泡。

I make hot bubbles.

"那怎么可能？你没命中，但还能抓到猎物？解释一下。"

"很简单，我的速度如此之快，以致产生热气泡。它把猎物弄晕了，这样我就得到晚餐了。"

"你一定是这里速度最快、最有力气的动物。我听说你移动的速度像飞行的子弹一样。"

"How does that work? You miss and still get your prey? Do explain."

"It is very simple. I move so fast that I make hot bubbles. These stun my prey, and there I have my dinner."

"You must be the fastest and the most powerful animal around. I heard you move with the speed of a flying bullet."

"是的，这是真的。你知道你永远做不到我做的这些事情，你不感到沮丧吗？"

"你可别小看我，虾女士！"

"你只是一条蛆，连腿和翅膀都没有，怎么还想和我比？"

"It is true, yes. Don't you ever become frustrated, knowing that you will never be able to do what I can do?"

"Don't you underestimate me, Ms Shrimp!"

"You are only a maggot. You don't even have legs, or wings. How would you even want to compare yourself with me?"

你只是一条蛆。

You are only a maggot.

······你还没一粒米大。

... no bigger than a grain of rice.

"我就是我，我为什么要像你一样呢？"

"但是你还没一粒米大，而且，我看你除了蠕动什么也不会。"

"我还会跳，而且跳得很远。"

"And why would I want to be like you, when I can be me?"
"But you are no bigger than a grain of rice, and all I can see you do is wriggle."
"I can jump too, and very far at that."

"别逗我笑了！你还会跳？谁信呢？我知道，那是不可能的。"

"请给我一个机会，我将证明你是错的。"

"我可没有这个耐心。不过，至少今天早上我笑得很开心，所以我得谢谢你。"

"你学过物理吗？"蛆问。

"听着，我是只虾，而不是科学家。我只需要知道如何抓住我的猎物，并且在这件事上我非常清楚该怎么做。"

"Don't make me laugh! You, jump? No one will believe that. I, for one, know that it is not possible."

"Please give me a chance and I will prove you wrong."

"I don't think I have the patience for that. But, at least, I had a good laugh this morning, so I must thank you for that."

"Did you ever study physics?" Maggot asks.

"Look, I am a shrimp, not a scientist. All I need to know is how to catch my prey, and I know very well how to do that."

别逗我笑了！

Don't make me laugh!

......我头上和屁股上的细小触须？

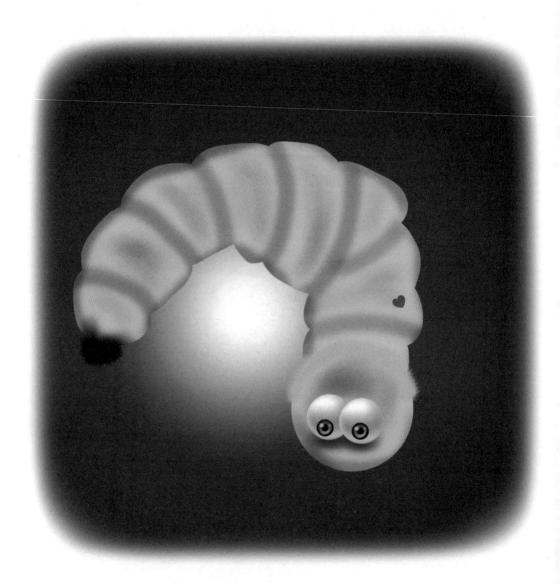

... little hairs on my head and my rear?

"而且我还知道如何从捕食者那里逃脱。"

"你知道？怎么逃脱？"

"你看见我头上和屁股上的细小触须了吗？"

"And I know how to escape from predators."

"You do? How?"

"You see these little hairs on my head and my rear?"

    "真不容易看见，不过，是的，我确实看到有两小簇。"

    "你知道吗？当我感受到威胁时，就把我的头和尾部拉到一起，这样这两部分就会接触并粘起来。就像小果实上黏糊糊的毛刺，你能理解吗？"

"They are really hard to see but, yes, I do see two little patches."
"When I feel threatened, I draw my head and tail together so these two patches touch and stick together. Like those sticky hair on a burr, you know?"

把我的头和尾部拉到一起。

I draw my head and tail together.

我就飞起来了。

I fly.

"当然。然后呢？"

"然后我就泵送体液，让自己肿胀起来。天哪，当这些黏性触须展开时，我就飞起来了。就像这样！"

"Sure. And then what?"
"Then I pump my body fluids so that I swell up, and girl, when those sticky hairs let go, I fly. Like this!"

"哇，看看你跳得多远。这就像魔法一样!"

"你或许认为这是魔法，而我则称它为软体机器人技术。这在我的日常生活中很普通……"

……这仅仅是开始！……

"Wow, just look how far you jumped. This is like magic!"
"Now what you may consider magic, I call soft robotics. And it is just normal in my ordinary life…"
... AND IT HAS ONLY JUST BEGUN!...

······ 这仅仅是开始！ ······

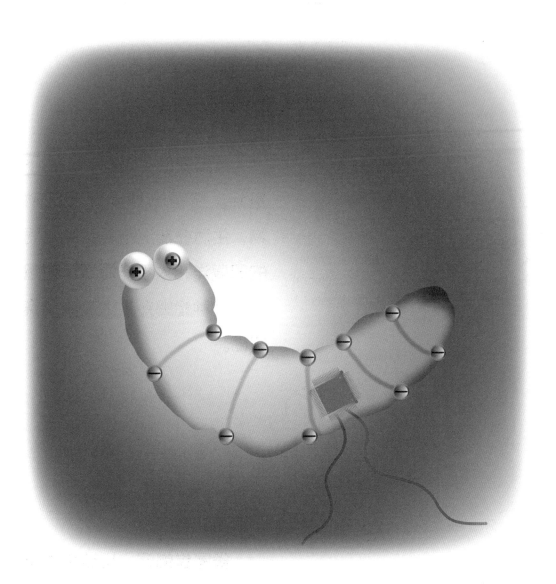

... AND IT HAS ONLY JUST BEGUN! ...

螳螂虾既不是螳螂，也不是虾。它是一种甲壳亚门动物，生活在非洲与夏威夷之间的印度洋和太平洋海域。有超过 450 种螳螂虾生活在珊瑚礁底部。

The mantis shrimp is neither a mantis nor a shrimp. It is a crustacean that lives in the Indian and Pacific Oceans, between Africa and Hawai'i. There are over 450 different species living at the base of coral reefs.

The peacock mantis shrimp, also known as the harlequin or clown shrimp, is one of the most colourful animals in the world. Its strength allows it to prey on animals much larger than itself.

雀尾螳螂虾，也被称为小丑虾，是世界上色彩最丰富的动物之一。它的力量使它能够捕食比自己大得多的动物。

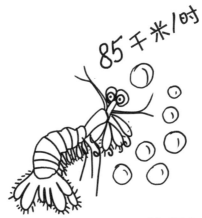

85 千米/时

螳螂虾以每小时 85 千米的速度"出拳"，能产生 1500 牛顿的力。这种力量非常强大，即使没有击中目标，它所产生的气泡也会使猎物眩晕。

The mantis shrimp's punch is delivered at a speed of 85 kilometres an hour, delivering a blow of 1,500 N (newton). The force is so powerful that even when it misses the target, the bubbles created by the movement will stun the prey.

螳螂虾用于击打的附肢上有一个特殊的减震器。它由带有小裂纹的弹性材料构成，这种裂纹结构是为了避免其不可思议的力量损坏自己的身体。

The mantis shrimp has a special shock absorber in its hitting clubs. It consists of elastic materials with small cracks that are structured in such a way as to avoid its incredible force damaging its own body.

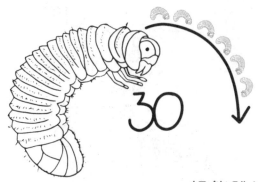

The maggot can jump up to 30 times its body length. Its fluid-filled hard bottom operates like a temporary leg, one that operates for just a brief moment and then reconfigures to the typical soft body that characterises the maggot.

蛆能跳出比它身体长 30 倍的距离。它那充满液体的坚硬底部可以临时当作腿来使用，但持续时间很短，然后就变回柔软状态，就像它身体其他部位一样柔软。

While traditional robotics uses hard, strong and sturdy materials, soft robotics uses soft materials, like those found in living organisms. These are mostly from natural sources, such as cloth, paper, fibres and rubber. These are lightweight, affordable and easily customised.

传统机器人技术使用坚硬牢固的材料，而软体机器人技术使用柔软的材料，如那些在生物体中发现的材料。这些材料源自大自然，如布料、纸张、纤维、橡胶等。这些产品重量轻，价格实惠，易于根据需要定制。

软体机器人通过承压流体等介质输入控制信号，集收缩、伸展、弯曲和扭转功能于一体。这在医疗应用中是必需的，如心脏辅助设备、柔软的机器人手套以及需要在不规则物体和表面上进行的工作。

Soft robotics combines contraction, extension, bending and twisting with control input such as pressurised fluids. This is required for medical applications like heart-assist devices, soft robotic gloves and the handling of irregular objects and surfaces.

软体机器人可以处理细小易碎或形状不规则的物体。这些机器人可以变形以适应不同的空间，克服了传统机器人刚性结构带来的安全风险。这种机器人集速度、力量和加速度于一体。

Soft robotics can handle thin and delicate objects as well as irregular shapes. These robots can adjust to space and overcome the safety hazards caused by the rigid construction of traditional robots. These robots combine speed, force and acceleration.

Would you like to work with robots that are hard and full of wires?

你喜欢和坚硬的装满线缆的机器人一起工作吗？

Is the tiny maggot, without any arms, legs or wings not just the most impressive little creature?

没有胳膊、腿和翅膀的微不足道的蛆，难道不是最令人印象深刻的小生物吗？

Would you like to be given a chance to prove what you can do, even when people say it is not possible?

即使别人说不可能，你也希望得到一个机会来证明你能做到吗？

What does a good laugh in the morning do for you for the rest of the day?

早上开怀大笑对你一整天的生活都有什么好处呢？

Do It Yourself!

自己动手！

We all know someone who may think that he or she is better than everyone else, let's see what we can learn from spending time with such a person. The first thing we should probably learn, is how not to act that way ourselves. When interacting with others, are you humble about your skills and abilities, or boastful, like the shrimp in the fable? Do you take the needs and feeling of others into consideration? By reflecting on questions like these, we can judge if we are as humble and considerate as we would expect others to be.

我们都认识一些自认为比别人优秀的人，看看我们能从与这样的人相处中吸取什么教训吧。第一条应该吸取的教训就是，我们自己如何避免以自大或者自恋方式行事。当你与他人交往时，你是谦虚地对待自己的技能和能力，还是像故事中的螳螂虾一样自吹自擂呢？你会考虑别人的需要和感受吗？通过反思这些问题，我们可以判断自己是否像我们期望别人做到的那样谦逊和体贴。

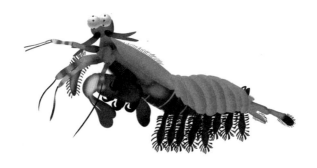

学科知识
Academic Knowledge

生物学	螳螂虾有复杂的视觉系统；螳螂虾的视网膜上有用于颜色分析的光感受器，以此在色彩斑斓的珊瑚礁中区分不同的事物；笑可以降低血压，改善心脏健康；尼龙搭扣是一种受植物启发的仿生无胶粘接方式。
化 学	以螳螂虾为灵感的碳纤维环氧树脂复合仿生材料；子弹的速度受推进剂燃烧速率影响；笑能释放内啡肽，这是一种天然的止痛药。
物 理	螳螂虾以刺击或打击的方式杀死猎物；猛击之后的气穴现象会在瞬间激发一道闪光和很高的温度；螳螂虾的附肢上有一层呈螺旋状排列的矿化纤维质，可以抗打击；螳螂虾的视觉系统能感知紫外线；声速和亚声速；力的单位：牛顿；螳螂虾能感知光的偏振。
工程学	作为制造机器人的材料，金属等传统硬质材料将被柔韧、有弹性的材料取代，以适应多变的环境；柔性材料工程为机器人技术和可穿戴设备引入了新方法，这些设备可以与人体配合，并能适应不可预测的环境；受螺旋形减震构造启发的航空航天器装甲和汽车装甲；螳螂虾并不是比较不同感受器感受的颜色，而是通过12个感受器将完整的颜色发送到大脑；步枪枪管内雕刻的螺旋形膛线能够使子弹在飞行中保持稳定。
经济学	与目前集中化和规模化的能源生产模式相对照的是地方化、小规模、分散的能源生产模式；新兴的生物计量学行业。
伦理学	为了自己的利益而忽视别人的需要。
历 史	据估计，螳螂虾有4亿年的历史，比恐龙早1.7亿年；在公元前400年的希腊已经有了小丑表演，那时的中国处于战国时期。
地 理	珊瑚礁是螳螂虾的家园。
数 学	柔性复合材料制成的机器人是非线性的；螳螂虾的视觉系统能接收4倍于人类的数据，同时由于其独特的色彩计算模式，它的大脑远小于人类大脑；理性是建立在以比值来表示的比较价值之上的。
生活方式	我们总是按大小、价格来进行比较，这是竞争型社会的基础，往往忽略了边缘人群的需要；脱口秀、连环画和卡通片——对笑的渴望。
社会学	媒体中的嘲笑是一种观点的延伸；傲慢使人与朋友、社群疏远；小丑的社会功能：提供笑声和娱乐，让人感觉更好，还能缓解慢性疼痛。
心理学	情商的力量，从自我意识、自我管理、动机、同理心到社交技能；贬低别人，表现出不尊重或过度傲慢；通过嘲笑别人来抬高自己；嘲笑是一种讽刺性的描述，它引导人们关注坏的或荒谬的事物，从而把人们的注意力从好的和合理的东西上引开；教育傲慢的人学会谦逊；谦逊幽默的用处；自信的人往往更成功。
系统论	自然界里，一切都像软体机器人一样运作，没有任何东西是靠电线或电池来运转的。

情感智慧
Emotional Intelligence

蛆

蛆对螳螂虾表现出的力量感到敬畏。他坦率地表达了对螳螂虾的钦佩，但并没有被螳螂虾压倒。他坚定而自信，有较强的自尊心。他以一种娴熟的方式让自己和螳螂虾处于平等的地位。他对自己很了解，并警告螳螂虾不要低估他。起初，蛆克制自己，但当螳螂虾看不起蛆的体形和能力时，蛆说自己能跳很远的距离。蛆礼貌地要求螳螂虾给他一个机会来证明这一点。螳螂虾提出更多问题，蛆平心静气地回答，在这个过程中，蛆赢得了螳螂虾的钦佩。

螳螂虾

螳螂虾知道自己独一无二的能力，并傲慢地吹嘘这些。她过于自信，用一种高傲的语气和蛆说话。她清楚地表示自己没有时间去关注一个没有四肢的小生物，并认为蛆自称能够跳起来很可笑，甚至笑出声来。尽管如此，受好奇心和怀疑心的驱使，她继续进行谈话。随着对话的展开，她变得不那么自大了，开始对蛆的能力表现出兴趣。她被蛆的表演所打动，并表达了她的感受，不由自主地展现了她的惊讶。

艺术
The Arts

螳螂虾的色彩感知能力是其他生物无法匹敌的，它的色彩展示能力也很少有其他生物能够企及。螳螂虾的色彩表现力，简直就像直接来自安迪·沃霍尔的油画！他常常以多种多样的色彩组合呈现同一主题。让我们看看利用无限的色彩组合进行创作是多么富有创意吧。画一幅你认识的、最喜欢的人的肖像。先画出轮廓，然后用不同颜色组合来上色，创作出系列图片。

思维拓展
Systems: Making the Connections

地球资源正在被过度开发，这威胁着地球上所有物种的生存。我们需要重构经济体系，使之成为一种能提高生活质量且遵循自然进化路径的经济体系。我们需要团结协作来实现目标，这就需要情商。情商对于人际关系至关重要。情商高的人能够理解他人的需求，与他人友好相处，从而形成有效的合作团队。情商高的人具有较强的自我管理能力，不诋毁对手，与他们友好相处，这能造就一个容易产生变革的平台。情商高的人尊重他人的观点，对他人的遭遇具有同理心，不固执成见，不吹毛求疵，这有助于创造一个富有效率和凝聚力的社会环境。技术方面，人工智能和机器人的使用将在以可持续发展为目标的社会转型中发挥关键作用，软体机器人也将如此。软体机器人技术是对传统机器人技术的一种革新。未来的实验室将充满微小的有机体，它们能撞击、咀嚼、跳跃，为我们提供在过去不可想象的技术方案。设计软体机器人时应当使用小的、基于自然灵感的机器人设计路径，这些机器人在力量和速度上接近它们的生物原型。自然界的许多生物所拥有的爆发力，并不是基于强壮的肌肉，而是依赖能像弓一样锁定和释放的弹性组织。强壮而有弹性的肌腱、角质层和其他弹性结构像弹弓一样伸展和释放，为跳跃和撕咬等动作提供动力，所有的机制都能增强力量。传统的性能数学模型与肌肉固有的机体平衡有关，肌肉可以更快速或者更有力地收缩，但不能同时做到既快速又有力。设计者需要考虑两者的内在平衡，而软体机器人技术正是一种能同时实现更快、更强、更有力的技术。让我们利用自己的情商以及大自然给我们的启迪来开发软体机器人，所有人都要为共同生活在这个星球上的所有物种的利益而携手努力。

动手能力
Capacity to Implement

软体机器人技术的灵感来自大自然。我们现在知道了螳螂虾和蛆的独特能力。除此之外，我们还可以加上长颈蚁和水螅的独特能力。现在，看看如何把这些独特能力应用到家用物品上。引导你的思维走向不同的科学研究领域，通过这样的方式，你甚至可能找到改变世界的创新思路！

故事灵感来自
This Fable Is Inspired by

希拉·帕特克
Sheila Patek

希拉·帕特克以优异的成绩获得哈佛大学生物学学士学位。随后,她在美国北卡罗来纳州达勒姆的杜克大学获得生物学博士学位。她拥有美国加利福尼亚大学伯克利分校的博士后身份。希拉给研究生和博士后科学家讲课。她的实验室为教师提供研究经历,使他们能够将他们的研究与新课程的开发相结合。她还帮助高中生在早期阶段探寻科学突破的机会。2019 年,她成为杜克大学生物系教授。在这个以她的名字命名的实验室里,她正在研究节肢动物的声学通信系统,以及螳螂虾和蚂蚁的动物运动力学。她是软体机器人领域的先驱之一。

图书在版编目（CIP）数据

冈特生态童书.第七辑：全36册：汉英对照 /
（比）冈特·鲍利著；（哥伦）凯瑟琳娜·巴赫绘；
何家振等译.—上海：上海远东出版社，2020
ISBN 978-7-5476-1671-0

Ⅰ.①冈… Ⅱ.①冈… ②凯… ③何… Ⅲ.①生态
环境－环境保护－儿童读物—汉英 Ⅳ.①X171.1-49

中国版本图书馆CIP数据核字（2020）第236911号

策　　划　张　蓉
责任编辑　祁东城
封面设计　魏　来李　廉

冈特生态童书

软体机器人

[比]冈特·鲍利　著
[哥伦]凯瑟琳娜·巴赫　绘

何家振　译

记得要和身边的小朋友分享环保知识哦！
八喜冰淇淋祝你成为环保小使者！